YOUR KNOWLEDGE HAS VALUE

- We will publish your bachelor's and
  master's thesis, essays and papers

- Your own eBook and book -
  sold worldwide in all relevant shops

- Earn money with each sale

Upload your text at www.GRIN.com
and publish for free

Stephen Gumboh

Promotion of Appropriate Agricultural Technologies and the Management of the Fertilizer Support Programme in Zambia

The Case of Small-Holder Farmers

GRIN Verlag

Imprint:

Copyright © 2012 GRIN Verlag GmbH
Druck und Bindung: Books on Demand GmbH, Norderstedt Germany
ISBN: 978-3-656-48107-2

This book at GRIN:

http://www.grin.com/en/e-book/231255/promotion-of-appropriate-agricultural-
technologies-and-the-management-of

Development Policy and Management

Promotion of Appropriate Agricultural Technologies and the Management of the Fertilizer

Support Programme in Zambia: The Case of Small – holder Farmers

Stephen Gumboh
Ministry of Finance
National Policy and Programme Implementation
Lusaka, Zambia

February 2012

Abstract

The paper explores measures put in place by the Zambian government over the years to enhance the small holder farmers' access to appropriate agricultural technologies and the management of the Fertilizer Support Programme. The paper focusses on the development of appropriate technology for small – holder farmers, who constitute the majority of the agricultural population in Zambia. Of significance are the analysis of government policy on the development on appropriate technologies and the management of the Fertilizer Support Programme to the advantage of small-holder farmers.

The paper finds the significance of small-holder farmers in economic development as a crucial factor. The paper further finds the government with no choice but to pursue the agriculture sector vision of an efficient, competitive, sustainable and export lead agriculture sector that assures food security and increased incomes. This entails the development of appropriate technologies and policy initiatives ideal to small holder farmers at every level.

Keywords: Development Policy, Agriculture, Appropriate Technology, Small-holder Farmers, Fertilizer Support Programme, Management, Mechanization Policy, Agriculture Research, Technology Policy.

Table of Contents

Abbreviations and Acronyms

CDT	Cotton Development Trust
CVRI	Central Veterinary Research Institute
FNDP	Fifth National Development Plan
FSP	Fertilizer Support Programme
FISP	Fertilizer Input Support Programme
GART	Golden Valley Agricultural Development Trust
GDP	Gross Domestic Product
GNP	Gross National Product
LDT	Livestock Development Trust
MALS	Ministry of Agriculture and Livestock
MRI	Maize Research Institute
NAP	National Agriculture Policy
NARDC	National Aquaculture Research and Development Centre
NARS	National Agricultural Research (& Development) Systems
NISIR	National Institute for Scientific and Industrial Research
PRSP	Poverty Reduction Strategy Paper
SCCI	Seed Control and Certification Institute
SNDP	Sixth National Development Plan
ZARI	Zambia Agriculture Research Institute

1 Introduction

1.1 Development and Poverty

Development is a concept that is pursued in varying forms in different economies. But the most pronounced is the one being experienced in developing economies. This is so because of the many economic and social problems that developing economies have been experiencing as a result of the political and economic changes that have continued to sweep through these economies over the years. The changes have fundamentally shifted the understanding of development in that a range of both new and unresolved challenges have continued to take the centre stage. Zambia is not an exception to these development challenges. One of the most pronounced challenge is that of the ever increasing levels of poverty and scanty development across the country.

The issue of persistent poverty has remained the major challenge in Zambia. This is because almost eighty percent of the population is locked in a vicious cycle of human suffering as a result of poverty. This is despite the economy registering consistent economic growth over the years. The situation of poverty is worse in the rural areas where most of the people are highly impoverished with no means of living decent lives due to lack of basic necessities of life. The poverty prone mainly suffers immensely from inadequate access to economic and social resources. The majority of the people in the rural areas are poverty prone because of their dependence on farming as the main source of livelihood. The problem is that such people are not able to even sustain themselves in their farming ventures as they lack access to improved seed, fertilizers, credit facilities and markets for their farm produce.

Poverty levels in rural areas are in contrast to those of urban areas, where some people have direct benefit from government's development initiatives. Such a situation is clearly evident in the ever increasing disparities between the rural and urban communities in the distribution of welfare. The disparities are attributable to the ever widening economic and social gap between the rich and the poorest in society. This is like history repeating itself when one reflects back to the era of colonialism, which has the root in the current development problems. As Worsley, (1984, p1) put it" the whole of history, is the history of development implying that the disparities being experienced in the present era have their root in the history of development and cannot therefore be completely overlooked". This could be the reason why the issue of poverty remains the Government's most pressing economic and social challenge.

1.2   Statement of the Problem

With poor prospects for copper and the limited opportunities for developing other potential industries in Zambia, the Government saw agriculture with its related agro-industries to be a potential growth sector in the economy and as a major effort to address the poverty issues mostly in the rural areas of the country. However, despite the agricultural sector having the potential for growth and development, "the sector has continued to experience low investment and low production and productivity especially among small-holder farmers of which 65 percent are women", (SNDP, 2011-2015, p.121). This is evidently clear when consideration is taken that of the twelve million hectares of arable land; only two million are continuously cropped to a greater extent. This is a clear reflection of the problem of production technologies available to farmers of which the majority are small- holder farmers who own or cultivate less than two hectares of land, mainly for their subsistence living.

Notwithstanding the substantial potential to contribute to the national food supply and to agricultural GDP, small-holder farmers are nonetheless found to constitute a third of the country's most poverty – stricken due to the unproductive nature of the type of agriculture in which they are engaged. The difficult is that most of the small-holder farmers have no appropriate technology and are only dependent on the traditional methods of production where hoes and cutlasses are used to cultivate the land. Further still, they have limited or no access to the supply of inputs such as fertilizers and other farming implements.

Lack of appropriate technology and access to other farm inputs are noted as major hurdles in the efforts of the small-holder farmers to enhance agricultural productivity. The understanding is that small-holder farmers' agricultural productivity is an important aspect to increased employment, higher rural incomes and the reduction in the disparity between rural and urban areas. However, technologies do not just fall like "manna from heaven". There are exogenously determined meaning that there are always other factors at play. Cost factors are some of the major determinant of technology. Cost factors, for example, are the major constraints to small-holders farmers because there are limiting factors to their quest for appropriate technology to carry out various farm initiatives.

1.3   Objectives of the Study

The essence of the study was therefore to explore the measures that the government had had to put in place over the years to enhance the small holder farmers' access to appropriate agricultural technologies and the management of the some interventions such as the fertilizer support programme (FSP).

Essentially, the study sought to examine the development of appropriate technology in relation to Zambia agricultural sector. In the discussion of development and appropriate technology, focus was on the small – holder farmers, who apparently constituted the majority of the agricultural population. An attempt was made to understand the appropriate technologies required by small-holders farmers to enhance their agricultural productivity. Of significance were the analysis of government policy on the development on appropriate technologies and the management of the FSP to the advantage of small-holder farmers. Further, the understanding of the requirements and constraints of the small-holder farmers were also of significance to enable the making of suggestions which were to guide government policy for a well integrated agricultural development and management programme.

1.4   Methodology of the Study

Due to limitations in the scope of the study and the non-availability of some statistical data on small-holder farmers in Zambia, it was not possible to do an empirical assessment of the development and adoption of appropriate agricultural technologies with regard to small-holder farmers in Zambia. The study only relied on the available documentation of the previous studies on agriculture in general and the situations that prevailed on the ground. Some selected discussions were nevertheless held with relevant stakeholders to confirm the accuracy of the information and possible outcomes.

2 Agriculture and Development – Review of Literature

2.1 Agriculture Terminology

Agriculture is synonymous with farming. It is often considered "as an art of science of cultivating the ground, including the harvesting of crops, and the rearing and management of livestock; tillage; husbandry and farming" (Brain Quote 2012). Cultivation in this sense means the production of food by preparing the land to grow crops either on a small or large scale. Husbandry, on the other hand has to do with the breeding and caring of farm animals. In farming, there are terms such as subsistence and commercial farming. Subsistence farming has to a great extent to do with the provision of the basic needs of the farmer without anything for marketing, while commercial faming mainly applies to situations where surpluses are available for marketing purposes. Subsistence farming is mainly practiced at a small scale and mostly in rural areas where the majority are small-holder farmers.

Growth in Agriculture often involves improvements by means of agricultural techniques. However, it is found that this growth, especially with small-holder farmers, is mainly constrained by inadequate extension services, high cost of financing, inadequate infrastructure and poor functioning agricultural marketing, among others.

2.2 Development Concept

In its simplest term, development would mean change. In the words of Thirlwall, (1995,p12)., development is " a process of transformation of the entire economic and social system which eventually lead to the improvement of the livelihood of the people and often follow a well-ordered sequence and exhibit common characteristics across countries". The sequence and exhibition of common characteristics across countries is however an issue of debate due to tremendous contrasts and diversities among different countries due to their physical and geographical positions which differentiate them by their economic status or the endowed characteristics. The debate is that it is not possible for developmental characteristics to be applicable to countries even if they are to follow a similar pattern because development is a process that does not occur haphazardly but is gradual and only reaches the level of full potential at later stages.

Development, in terms of social and economic phenomena, goes beyond the traditional paradigm of development, where it is linked to the growth of the Gross National Product (GNP). The GNP traditional understanding of development mainly looked at the rapid gains from the increase in GNP in an economy and their "a trickle-down effect. The understanding was that "through these gains, the masses would eventually benefit through the creation of the necessary conditions for equal distribution of economic and social benefits of growth" (Todaro, 1997, p 14). This was more of a materialistic notion of development were issues of output, employment, and income per capita were more pronounced than the beneficiaries or the changes in human conditions. Changes in human conditions are mainly associated to poverty, unemployment and inequality in the distribution of welfare. In the new development thinking, the issues of poverty and its effects on the lives of people seem to take the centre stage in the development focus. The understanding is that higher levels of poverty in any country have greater effects on the economy in that they completely over-shadow any economic achievements wherever they persist. Development in such an environment would mean more than an attempt to reduce poverty in order to achieve social, economic and political improvement for the betterment of society. This is because the lack of development in an environment is usually characterized by poverty, hunger and disease which undermine the human potential to develop. The presence of poverty, hunger and disease in an environment eventually lead to political conflicts. The issue of development therefore becomes crucial in that development problems perhaps more than the problems associated with any other aspect of human life, are the problems of policy. It is an issue of self-determination, which comes into play when people consider the benefits that accrue with development.

2.3 Technology for Agriculture Development

2.3.1 Transferred Technology

The concept of transferred technology mainly linked to modernization theory which entailed the transfer of capital intensive technology from advanced countries to developing countries. According to Lewis, (1954, p15), "the transfer of technology in the traditional sector occupied a key position in the modernization thinking, because it was seen as presenting the opportunity to accelerate the transformation of traditional agriculture to modern agriculture". The conventional view regarded the transfer of technologies mainly through genetic material (for instance, high yielding varieties of rice and wheat) or machines (such as tractors) as well as chemical innovations (fertilizers, pesticides, etc).

However, since the characteristics of the developing countries were generally held to be labour surplus and capital deficient, the imported transferred technologies were found to be inappropriate for these countries and could not therefore be transferred. Worse still, to most farmers, especially the small-holder farmers, the transferred technology proved to be costly, more often than not. Given the differences in social – economic conditions between and also within countries, the transfer of technology had more often than not, failed to yield the expected results as there were found to be inappropriate.

2.3.2 Case of Appropriate Agricultural Technologies

In the analysis of appropriate technology, it is noted that the first and obvious determinant of a country's appropriate technology is "its factor endowments – that is, the relative proportion in which labour, capital, land, skills and natural resources are available to an economy" (Singer, 1977, p1). But it is important not to be wary of going to the extremes in thinking that if a country is labour or capital abundant, then it should just use the relatively abundant factor. There is the possibility that a country which suffers from a relatively shortage of capital and is abundant in unskilled labour may choose a technology in a given sector or a particular project which uses more capital and less skilled labour. The underlying factor is its appropriateness in terms of reliability, affordability, sustainability and above all, productivity. If a technology has these factors, it could then be considered appropriate.

Loosely, it can be understood that the technology used in the informal sector would often be nearer to an "appropriate technology", than that used in the modern sector. However, when looked at critically, both could be inappropriate in the sense that one uses too much capital relative to what the factor endowment would permit, while the other uses too little capital (Singer, 1977, p1). In strict sense, it could be understood that the technology required for transforming the agriculture systems would be an intermediate technology – one which could be between the high capital intensive technologies of the modern sector and the indigenous technologies of the traditional sector. Dickson (1974,p154), considered intermediate technology as "a set of technologies lying mid-way between the capital intensive technologies usually exported to the under-developed countries by the industrialized nations and the indigenous technologies that under-developed countries possess".

There is however a caution on the use of the term "intermediate technology". This is because even if the concept makes sense to a certain extent, the terms "midway" and "exported" in the definition pre-supposes intermediate technology to be a technology the industrialized nations have already had and is sort of outdated. The implication is that having advanced beyond such technology, the only possible way to get rid of it is by transferring it to the developing countries where it could be regarded as new in its application. By accepting such a technology, it would always suggest that developing countries would forever follow in the footsteps of the developed nations technological advancements.

Considering that different countries and regions develop at different paces in accordance with differences factor endowments and social – economic environment, the concept of appropriate technology utmost supposes a technology that is affordable and is able to yield sustainable and increased productivity in particular environments. It matters less, whether the technology is biased towards the labour or capital factor, so long as it yielded the expected results and was accessible to a large sector of the community. In other words, appropriate technologies under prevailing conditions in developing countries are expected to be those that are more productive and are often highly labour intensive, less costly and more manageable than large scale labour saving and capital – intensive technologies of the industrialized countries.

2.3.3 Appropriate Technology Concept

Probable, the most comprehensive explanation of "Appropriate technology" is advanced by Norman and Hays (1979, p.68), where four basic elements are listed to constitute appropriate technology. To them, appropriate technology has to be technically and economically feasible, socially acceptable and infrastructural compatible.

As for technical feasibility, this has to do with the productivity, given the available technical elements in a particular production environment. If the technology is well- able to increase productivity, given the existing technical elements, then it is technically feasible. Economic feasibility, on the other hand, has to do with the dependability and compatibility of the technology with the farming systems. Essentially, the systems have to be profitable to the farmer. Further, technology has to have the flexibility to be easily reproducible with the minimum external assistance. It also has to have the ability to meet the basic needs without exerting excessive demands on the savings and foreign exchange resources of the country. In terms of social acceptability, the technology must be in line with the social structures, norms and beliefs of society. As for infrastructural compatibility, it is the level of infrastructure that is of utmost importance implying that technology must have requirements which could be easily accommodated by the present level of infrastructure.

Appropriate technology must therefore be viewed as a set of techniques that allows optimum use of available resources in a given environment. This means that technology should be consistent with the level of development of the production environment in general and of the farmer in particular. The level of development of the farmer would refer to the farmer's technological know-how and his ability to comprehend, appreciate and use the new technology. There is no rationality, for instance, to advocate tractors for subsistence farmers who even if they could afford, have no knowledge of how to use them, let alone maintaining them.

3   Agriculture Development in Zambia

3.1   Technology Development

The technological development in Zambia dates back to era of economic reforms where the role of agriculture was clearly spelt out as one important catalyst to economic diversification from copper driven economy. The Government regarded agriculture as the provider of raw materials for import substitution industries in the sense that it was perceived as an essential sector to increase peasant incomes thereby raising their purchasing power and at the same time expanding the consumer market for the industrial sector. This was to further enhance the generation of wage employment in rural areas of the country.

3.2   Mechanization Policy

In its quest for development in agriculture, the government put forth the mechanization policy. The major emphasis in mechanization policy was the wholesome introduction of tractor mechanization in Agriculture settlement schemes. Essentially the idea behind the creation of agriculture schemes was to make easy the provision of not only high technology equipment, but also communal water supplies, clinics, schools and other services within the framework of rural development.

To show commitment to mechanization policy, Government managed to distribute tractors to all agriculture areas in Zambia. Remarkably, nearly all the peasant farmers did appreciate the role of technology in the production process although their reactions varied according to their financial power and access to private or state infrastructures and credit facilities. Notably the mechanization policy was accompanied by a loan facility. The idea behind this loan facility was to make possible the purchasing of new productive technology, especially through the family farm resettlement schemes.

The mechanization policy however had its own problems. One such problem had to do with the hastiness in which agricultural schemes were created without proper and adequate planning. Equally overtime, most of the tractors in the schemes were grounded and could not be repaired as most of the scheme members did not have any basic engineering training to repair even simple faults. In order to overcome the lack of skilled manpower in the schemes, the government sourced for expert knowledge to the development of the agriculture schemes into cooperative movements (Chipungu, 1988, p.168). In this capacity building programme, the need for new techniques of production was emphasized. The problem however came in the provision of modern equipment of which the cooperative members had no knowledge about, not to mention the modern agricultural practices. Instead of being a solution to the overriding problems of mechanization, the approach tended to aggravate the problems. The situation made it even difficult for the introduced technology to be adopted and diffused in the process since most of the activities could not be transferred to the locals due to lack of skills and knowledge among cooperative members and incompetent managers and extension officers to effectively exploit the new production techniques. Lack of management skills manifested as the weakest link almost everywhere in the adoption and diffusion of the new production technology. Like many donor driven projects, the mechanization programme lacked continuity. The failure of mechanization policy led to rapid policy changes as faults were being discovered.

3.3   Reform of Technology Policy

In view of the disappointing performance of the small-holder/peasant farmers, given the availability of the modern agricultural equipment, both the government and the private credit offering institutions become reluctant to finance what was considered complicated technology for the small-holder/peasant farmers. This is what brought in the notion of ox-driven technology that could be cheaply acquired and operated by family labour rather than tractor and accompanying gear. On the part of the government, the need for reform of the technology policy was "to exploit to the fullest those resources which were are available in Zambia, that is, animal power, simple but effective machinery, etc. and then to develop the skills and knowledge of the indigenous people before thinking of sophisticated and expensive machinery which only the selected few were able to use" (Chipungu, 1988, p.168).

The government's drive towards animal power instead of sophisticated modern equipment was supported by local experts in farm machinery. They discredited fuel –powered implements on account of the high cost of acquiring them and the difficulties with the usage of spare parts and presumed absence of repair facilities especially in the rural areas. Animal power was presented as being sustainable and affordable. Given the emphasis on animal power, imports of ox-drawn implements such as ploughs, cultivation shellers etc, recorded an increase in importation. Despite the increase of the ox-drawn implements, fuel- powered agricultural equipment also continued to come into the country through the efforts of the government and private institutions to accommodate the demands of the few commercial and other emergent farmers who found these fuel powered equipments as appropriate.

4   Appropriate Agricultural Technologies for Small-holder Farmers

4.1   Problem of Transferred Technology

In Zambia, like in many other developing countries, there had been a tendency to transfer the imported technology to rural areas where the majorities are small-holder farmers. Because of the lack of proper planning and institutions in terms of adoption and diffusion of the imported technology, evidence indicated that the development of small-holder agricultural sector had not really come forth despite having the new productive technology. Given the non availability of technical elements in the socio-economic environment in which these technologies were adopted, no sustainable productivity was found in many of the cases. The argument was that the new imported technologies were infeasible, both technologically and economically. This was attributed to the technologies non dependability and compatibility with the existing farming systems.

The other problem found with imported technologies was the aspect of location – specific. Given the existing institutional capacity in Zambia, this made it difficult for the imported technologies to be found beneficial to the majority of the small-holder farmers. The problem of location- specific meant that since these agriculture technologies were developed according to the available institutional capacity at the original site of location, when transferred to Zambia, most especially to the small –holder farmers in rural areas, they were found to have limitations in their use.

There was also the issue of discrimination among small-holder farmers mostly exhibited by the traditional institutional system that tended to favour certain groups of small–holder farmers in terms of inputs distribution. For instance, where a new technology such as the high yielding varieties which requires information, credit access to irrigation and so on was introduced, given the structure of the receiving institutions, it was only the strong and elite small-holder farmers with the necessary information and credit facilities who took full advantages of the capabilities of the new seed varieties. With other small- holder farmers, the problem was that there were not in a position to take advantage of the high yielding varieties because they demanded applications of huge quantities of fertilizer to which most small - holder farmers could not afford due to lack of access to financial resources to enable them to acquire the required inputs. Even if the access to credit facilities were available, the cost of borrowing was usually beyond the reach of most small-holder farmers.

4.2   Appropriate Technology Policy

Appropriate technology is regarded by most agriculture proponents as one critical factor to increased agricultural production and utilization in almost all economies of the world. The Zambian National Agricultural Policy (NAP) clearly outlines its main objective which is "to develop a vibrant, competitive and efficient agricultural sector which is able to significantly contribute to income and employment generation, increased industrial development, export earnings and overall economic growth and poverty reduction"(NAP,2009,P11). Within the NAP, the development and promotion of appropriate technologies are strategically focused on appropriate farm machinery, implements, equipment and accessories and fishing methods. Others are the development of appropriate seed varieties and planting materials, livestock types and breeds; suitable fish species and aquaculture practices; pest and disease control; sustainable yield enhancing farming methodologies and appropriate use of biotechnology.

Other than the above, there are other facets of the agricultural sector in the policy such as the production of food and cash crops, inputs, agro-processing, agricultural marketing including exports and sustainable resource use. Others are livestock and fisheries development, irrigation, biodiversity, agricultural research and extension services, co-operatives and farmer organizations and institutional and legislative arrangements. They also include crosscutting issues such as HIV/AIDS, gender and the environment.

4.3   Agricultural Research and Development

As for research and development of appropriate technologies at country level, government took measures to increase production in the small scale sector in order to raise the standard of living of most of the people specifically in the rural areas. This was government's main focus on the agricultural sector in all of its national development plans. However, the attainment of sustainable increased production was not tenable without appropriate agricultural technologies. Appropriate technologies cannot be attainable in the absence of appropriate institutions to effect and support this technology. This is because appropriate agricultural technologies can only be generated through internal research for easy adaption to specific situations either by diffusion or extension training.

Agriculture research is found to be essential as a learning process on how to overcome, partially or entirely the limitations imposed by the natural environment and how to make the most effective use of the potentially favorable resources. As Mellon, (1969,p.441), put it "the first step in an agriculture development programme should be in the initiation of a substantial highly integrated research programme, directly connected to farm problems at one end to basic research and foreign effects at the other". This is the reason why agriculture research as an activity cannot be easily ignored by the developing economies. This is because through research, a country is given the ability to modify its environment to have a high level of agriculture production. What is important is for agricultural researchers to identify the many economic, human and social constraints which are likely impediments to the adoption of new practices among the small-holder farmers and then devise solutions to them. The constraints may include low incomes and discriminatory credit facilities in favour of few large commercial farmers at the expense of the majority small-holder farmers. The ultimate goal of research is to lead the development of appropriate technologies by effecting technical innovations in the production process so as to enhance productivity at the level of small-holder farmers.

4.4   Fertilizer Support Programme

Another Government initiative for small-holder farmers was the Fertilizer Support Programme (FSP). The programme was initially set up in 2002 as a response to the Country's Poverty Reduction Strategy Paper (PRSP). This was with the realization that failure of the agricultural sector to provide livelihoods for the majority of the people in rural areas attributed to persistent rural poverty. The FSP was therefore one broad reform in the agricultural sector. Its introduction was mainly to stimulate growth in the sector and improve its performance to reduce poverty and enhance food security in the country. The FSP involved supplying fertilizers and seed at a fifty per cent subsidy (PRSP, 2002, p.4).

Generally, the FSP objectives were to promote the use of low input and conservation farming technologies among selected target small-scale farmers who meet the criteria and to rebuild the resource base of the small-holder farmers. Conservation farming, to some stakeholders, was found to be the best method if exploited could produce maximum yields and aid long term soil and water management. The argument was that conservation farming method help in minimizing air, water and soil pollution. Practically, it was argued that some farmers who had been engaged in conservation agriculture were found to have produced enough food for consumption and excess for commercial purpose without negatively affecting the environment. This therefore made the method to be perceived as cost effective and very helpful to farmers to implement sustainable farming practices. According to Dr Mupangwa (as quoted in the Times of Zambia of 14th February 2012), "Conservation agriculture method was the best method to be used and applied by farmers because plant nutrients, bacteria, sediment and crop chemicals can be mitigated so that pollution of ground and surface water is curtailed". His argument was that once the usage of conservation agriculture is applied, it could eradicate the challenges facing farmers in the acquisition of farm inputs.

Other than the above, the FSP programme specific objectives hinged mainly on the increase of private sector participation in supply of agricultural inputs to small-holder farmers and the timely, effective and adequate supply of agricultural inputs across the country. Others were improved access of small-holder farmers to agricultural inputs; comprehensiveness and transparency in the distribution of inputs and risk-sharing mechanism for small-holders farmers. The expansion of markets for private sector supplies and increased involvement in the distribution of agricultural inputs in rural areas and facilitation of the process of farmer organisation, dissemination of knowledge and creation of other rural institutions to contribute to the development of agricultural sector were other specific objectives of the programme.

5   Study Findings and Comments

5.1   Role of Small- Holder Farmers in Agriculture Development

The study found that small- holder farmers were important players in Zambia's agriculture sector. Their contribution to people's livelihoods and to economic development was very evident in most areas where the majority was based. However, their contribution was found to be greatly affected by some of the economic policies which have had a greater effect on their productivity. The many constraints that the small-holder farmers were experiencing allowed only a few small-holder farmers to be actively involved in productive farming whilst the rest were just left to continue in subsistence living or in abject poverty. Again those small-holder farmers who were still viable enough had to be subjected to minimal investment in the areas of their agriculture activities making them even less competitiveness. This was evident in the decline of crop yields over the years. This was however attributed to unproductive methods of farming which had remained traditional. The decline in productivity and farming profitability in rural areas caused a number of the majority small-farmers holders to abandon their farms and migrate to the urban areas to search for other forms of livelihood further causing a strain on the problem of food security in the country.

5.2   Appropriate Development Technologies for Small-holder Farmers

In line with the country's policy on agricultural research, a number of research institutions were established by government. Notable among them were the Zambia Agriculture Research Institute (ZARI); Seed Control and Certification Institute (SCCI); Central Veterinary Research Institute (CVRI); National Aquaculture Research and Development Centre (NARDC) and the National Institute for Scientific and Industrial Research (NISIR). These institutions research focus were mainly on soils and crops, plant protection and farming systems, seed systems development; animal production, animal health and livestock related biotechnology; fish breeding, hatching and genetic conservation.

Apart from the government institutions, other private or joint shareholding companies complimented government agricultural efforts such as the Maize Research Institute (MRI) and ZamSeed. The MRI's focuses mainly on the development and seed distribution of the maize varieties while ZamSeed has the mandate to carry out development research on maize seed variety. Some minor research on other crops such as wheat, sorghum, sunflower, pearl millet, soybean, groundnut, Irish potato, pasture and vegetable seed among others are equally done. Seed-Co, a private seed company, is involved in the production and marketing of seed maize and has an established research centre for maize variety development.

As for Livestock development, the Livestock Development Trust (LDT), a private-public partnership organization was established to promote environmentally friendly livestock development initiatives for all livestock farmers in a gender sensitive and participatory manner. Equally, LDT allowed livestock farmers to attain profitable livestock-led farming systems that were to improve productivity and add value to raw production through transfer of skills, information and technology, and improving credit and market access. There is also the Golden Valley Agricultural Research Trust (GART), a partnership venture of the Government of the Republic of Zambia and the Zambia National Farmers Union (ZNFU), established to optimize the production, commerce and trade of crops, milk, chicken, goats and where possible their by-products, and income security of the target beneficiaries through Integrated Agricultural Research for Development programmes for market-oriented small, medium and large scale male farmers. Then there is the Cotton Development Trust (CDT), established for the purpose of conducting cotton research through the support of private and public sector funding.

Despite these research efforts, evidence suggested that the research findings were not beneficial to most of the farmers especially the small holder farmers who constituted the majority of the farming community as only few commercial and emergent farmers were found to be the main beneficiaries of the most research carried out by research institutions. The finding suggested the inappropriateness of research findings to the disadvantage of majority small-holder farmers. This was because research focus did not take into account the basic needs of the intended group of farmers. The other thing was that despite the development of appropriate technology depending to a large extent on a good knowledge and understanding of the existing production environment and related factors surrounding the individual farmer, in many cases, the research conducted had not been so cohesive to bring forth the intended results. The problem however laid on the fragmentation of agricultural research among a series of specialist disciplines coupled further with the packaging of innovations which were not designed as a whole. Further, there were also the diversions of researchers from the primary purpose of the research due to intellectual attractions of pure science.

Another notable aspect was the conduction of most of the research without accompanying extension services. This is because once technology has been generated through research it has to be diffused through extension services. Such a technology becomes appropriate and profitable at the time it is adopted by the farmer. It was however found that efforts to undertake extension services were greatly beset by a number of problems bordering on limited financial resources and transport facilities, coupled with lack of supporting staff and a severe shortage of qualified staff. The problem made it difficult for the new agricultural practices to be practically diffused or adopted by most of the small-holder farmers.

5.3   Implementation of the Fertilizer Support Programme (FSP)

An analysis of the FSP revealed a number of challenges in its implementation. The general view was that FSP since its inception had not performed to its expectation. The view was that the FSP had not made a great impact on the food security and poverty reduction especially in the poverty stricken rural areas. A number of research findings brought out several factors that contributed to the programme's failure to meet its objectives. Some of the identified factors hinged on inconsistencies in the supply of inputs and the mismatch in the distribution of fertilizers and seed. Other than these, there were also delays in inputs supply; poor transport facilities; inadequate supply of farm inputs and poor marketing arrangements. Delays in payments to farmers for the farm produce during the marketing season and the lack of non-use satellite depots were other cited factors. There were also issues of high inputs prices and low prices for farm produce and lack of monitoring and evaluation of the programme. Despite the factors cited, in the first seven years of FSP implementation, the programme was found to have recorded a significant improvement in the accessibility of agricultural inputs (i.e. fertilizers and improved maize seeds) by small- holder farmers.

Statistically, in the period from its inception in 2002 up to 2009, a total of 422,000 Mt of fertilizer had been distributed to small holder farmers, covering a total of 1,505,000 hectares of small scale maize. On an annual basis, an average of 60,000 metric tonnes of fertilizer covering about 150,000 small scale farmers, (each with a 1 hectare input pack for maize) were supplied countrywide.

The table below presents the breakdown of FSP fertilizer distribution in the period 2002 - 2009 periods:

Table 1: FSP Performance since Inception

Season	Fertilizer Amount	Number of Farmers
2002/03	48,000	120,000
2003/04	60,000	150,000
2004/05	50,000	125,000
2005/06	50,000	125,000
2006/07	84,000	210,000
2007/08	50,000	125,000
2008/09	80,000	200,000
TOTAL	422,000	1,505,000

Source: Ministry of Agriculture and Livestock

Despite the above positive results, a number of concerns about the FSP performance in the stated periods were raised by stakeholders. Most of the concerns hinged on the targeting approach of the FSP's beneficiaries and its impact on household and national food security. Other concerns were on the effect on private sector investment and participation in agricultural inputs supply markets and the programme's long-term sustainability, given the ever increasing competition for national resources by various sectors.

Specifically, most of the stakeholders raised concerns on the performance of the FSP in areas such as the poor targeting of farmers/beneficiaries and delays in input distribution. There were also issues on poor fertilizer use efficiency among targeted farmers resulting from inconsistencies in policy implementation, especially in the reversal of plans to reduce the subsidy level, and to stimulate agro-dealer development. Poor monitoring of program effects was another factor that made it difficult to sort out program achievements against objectives.

Arising from stakeholder concerns, the government in 2009 decided to restructure the FSP with a view of improving its operational efficiency and expanding the sphere of support to the farming community. As part of the restructuring, the programme was renamed Farmer Input Support Programme (FISP). The new programme incorporated camp agricultural committees to assist in the identification and selection of the target beneficiaries'. However, the programme experienced a reduction of the input pack size from 8 x 50kg bags of fertilizer and 1x 20kg bag of maize seed to 4 x 50kg bags of fertilizer and 1x 10 kg bag of maize seed (Ministry of Agriculture and Livestock, 2009).

For 2009/10 season, a total of 106, 836 metric tonnes of fertilizer and 5,342 metric tonnes of maize seed were distributed to 534,000 small-holder farmers. In the 2010/11 season the number of beneficiaries' rose to 891, 000 small- holder farmers with a total of 8,790 metric tonnes of certified seed and 178,000 metric tonnes of fertilizers distributed. As an additional incentive, the programme also distributed 30 metric tonnes of upland rice to promote crop diversification and food security in rural households (FNDP 2011 Annual review). Despite the reduction in input supplies, the country was able to achieve a bumper harvest of close to 2.8 million metric tonnes of maize grain in the 2009/2010 season. The success was attributed to the reforms undertaken by government and improved extension services. Despite the FSP being in place, there were still concerns from stakeholders to have it reformed further. This had to do with the funding of the same farmers with fertilizer and seed and then making them to graduate from the programme so that their farming becomes proper businesses than just self-sustaining. The suggestion was to have a revolving fund and not to allow the programme to be manipulated at will.

5.4   Options for Government

Having considered the many issues to do with agricultural productivity among small-holder farmers through the provision of appropriate technologies and the fertilizer support programme, there were still remain issues to be addressed in as far as the reduction of hunger and rural poverty in most areas of the country. Also there was this possibility of government to increase the productivity of those small-holder farms so as to allow them and the country at large to escape from the hunger and poverty.

The study findings proved the possibility of the government making radical efforts to enhance small-holder farmers' productivity and livelihood by empowering them with the means to access the crucial production resources. These resources were to range from land accessibility, water supply, energy provision, and credit facilities. The provision of appropriate technologies could further provide opportunities to small-holder farmers to develop their skills and to access the information on how to use them. There were also the issues of proper markets for products and inputs and good health care and sanitation facilities together with education and social services.

As for extension services, the government had no option but to take a remarkable approach to draw heavily upon local resources, values and skills to undertake such an important agricultural activity. There was also the need for government to pump in more resources into the research institutions and to other related extension services. Heavy investment in agricultural research and extension services was found as a major factor to the remarkable progress in the development of agriculture in many developing countries.

As for small-holder farmers, the government needed to take serious consideration of the role of small-holders farmers in agriculture development. However, the concern for development of small – holder sector farmers did not only call for the funding of technologies which were peasant – biased, but also called for hard work in attempting to analyse, develop and promote, for specific situations, institutional changes which would re-enforce their technological efforts. It was in this vain that appropriate institutions were seen more visibly at the local level and suggested that "meaningful rural development required the development of appropriate technologies that would suit local conditions, technologies cheap enough to be accessible to a larger sector of the community without excessive demands on the foreign resources of the country" (Schumpeter, 1973, p.69).

A major challenge however, for Zambia, in the development and adoption of appropriate technology was to develop an effective scientific and institutional capacity to design and adopt location specific agricultural technology, according to the resource endowments and socio-economic environment in which the new agricultural technology was to be employed. The failure of the government to institutionalize domestic research capacity had serious impediments to the development and diffusion of new technologies. All in all, the appropriateness of technology was in effect its adaptability to local conditions, its sustainability for assimilation in existing socio-economic conditions particularly with regard to small holder farmers and its acceptability by existing workers to handle it and its capacity to utilize local input.

As for development, there was a general agreement that any development which had an effect on the concerned population needed to be consensually accepted. This was achievable in situations where population groups in society were allowed to participate equally in the development process. In other words, there should be people's self-determination in the decisions concerning their welfare and the management of their own economic affairs. It was agreeable that development should be integrated and not of the type which benefited one sector of society and adversely affected another. It was viewed that development was much more than an economic or social process but a process which needed to be in harmony with the environment and dedicated to the well being of society.

The other crucial aspect of development was human potential. This was crucial in that it linked to the underlying concerns of many developing countries and that was the eradication of poverty and pursuit of equity and identification of tangible solutions to human development problems. Human development in a social perspective was ideal in addressing the problems of access to resources, provision of the basic needs and the distribution and effective use of scare resources. According to Todaro (1997,p18), "development in all societies had to have at least three objectives; to increase the availability and widening the distribution of basic life-sustaining goods; to rise the levels of the living standards of the people and to expand the range of economic and social choices available to the individuals". Anand and Sen (1995), in the 1995 UNDP Human Development Report, equally presented a number of areas which needed attention to advance human development. These hinged on "higher average life expectancy, lower infant and child mortality rates, higher literacy and a higher value of human development index as measures of development".

6   Conclusion

The way forward for Zambia hinged on giving priority to poverty and development challenges if it was to fulfill its development strategies. The understanding was that the kind of policies and issues needed to address poverty and other related issues of unemployment, are almost the same policies that are required to make economic growth and development both viable and sustainable. The role of small-holder farmers in poverty reduction and development in the rural areas could not therefore be underplayed by government. As Delgado (quoted in Chipokolo, 2006) put it, 'small holder agriculture is simply too important to employment, human welfare and political stability to be either ignored or treated as just another small adjusting sector of a market economy.' In other words, the sector was found to be the core to an agrarian economy like that of Zambia.

The issues constituting and surrounding the problem of small- holder farmers might appear complex and certainly intractable but were manageable and soluble. The country only needed government action at the level of policy implementation, an assertive and committed role at the level of the professional and policy maker and a freer and a more decisive participation at the level of small-holder farmer. The significance of small-holder farmers in economic development was found crucial and could therefore not be over-emphasized. The government was found with no choice but to pursue the agriculture sector vision of "an efficient, competitive, sustainable and export lead agriculture sector that assures food security and increased incomes by 2030 with the goal to increase and diversify agriculture production and productivity so as to raise the share of its contribution to 20 percent of GDP" (SNDP, 2011). As for small-holder farmers, it was found advantageous to for them to benefit greatly on the perceived enhancement of research and extension services and promotion of improved seed varieties as envisaged in the strategic focus in the SNDP. This was important for enhanced agriculture productivity.

Bibliography

Anand, S and Sen, A (1995), Sustainable Human Development Concepts and Priorities, UNDP Report, New York

Arnon, L (1981), Modernization of Agriculture in Developing Countries: Resources, Potentials and Problems, John Wiley and Sons, New York

Chipungu, S.N(1988), The State ,Technology and Peasant Differentiation in Zambia: A case study of the Southern Province 1930-1986, Historical Association of Zambia, Lusaka

Chipokolo, C (2006), Small-Holder Agriculture: Ignored Gold Mine, PELUM Zambia Policy Brief – Issue No. 1, SARPN, Lusaka

Civil Society for Poverty Reduction (2005), Targeting Small Farmers in the Implementation of Zambia's Poverty Reduction Strategy Paper (PRSP), Southern African Regional Poverty Network, Lusaka

Dickson, D (1974), Alternative Technology: Politics of Technical Change, William Collins and Co. Ltd, Glasgow

Government of the Republic of Zambia (2011), FNDP 2010 Annual Review, Ministry of Finance and National Planning, Lusaka

Government of the Republic of Zambia (2006), 2006 -2010 Fifth National Development Plan (FNDP), Ministry of Finance and National Planning, Lusaka

Government of the Republic of Zambia (2011), 2011 -2015 Sixth National Development Plan (SNDP), Ministry of Finance and National Planning, Lusaka

Government of the Republic of Zambia (2008), FNDP 2007 Annual Review, Ministry of Finance and National Planning, Lusaka

Government of the Republic of Zambia (2011), Fifth National Development Plan (FNDP) 2010 Annual Review, Ministry of Finance and National Planning, Lusaka

Government of the Republic of Zambia (2002), Poverty Reduction Strategy Paper, Ministry of Finance and National Planning, Lusaka

Government of the Republic of Zambia, (2009), National Agriculture Policy, Ministry of Agriculture and Livestock, Lusaka

Koloko, E.M (1979), International Technology for Economic Development: Problems of Implementation, in B.Turok .ed, Development in Zambia, Zed Press, London

Lewis, A.W, (1954), Economic Development with Limited Supplies of Labour, Manchester School of Economic and Social Studies, Manchester

Meier, G.M and James, E.R (2007), Leading Issues in Economic Development, Oxford University Press, London

Ministry of Agriculture and Livestock, (2009), Ministerial Statement to Parliament on the Performance of the Fertilizer Support Programme, Lusaka

Norman, D.W and Hays, H (1979), Developing a suitable Technology for small Farmer in National Development vol.21 No.3

Schumpeter, E.F. (1973), Small is Beautiful: A study of the economy as if people mattered, Blond and Biggs, London

Singer, H (1977), Technologies for Basis Needs, International Labour Organisation (ILO) Geneva

Snyder, M (1995), Transforming Development, Intermediate Technology, London

Thirlwall, A (1995), Growth and Development.6th ed, Macmillan, London

Thirlwall, A (2006), Growth and Development: With special reference to Developing Economies, 8th ed. Houndmills, Basingstoke, Palgrave Macmillan, New York

The Post Newspaper, 14th February, 2012

Times of Zambia, 14th February, 2012

Todaro, M.P (1999), Economic Development in the Third World, 6[th]ed, Oxford University Press, Oxford

Worsely, P (1984), The Three Worlds: Culture and World Development, Weidenfeld and Nicolson, London